Astronomer's Journal

This journal belongs to:

Name: _____

Contact: _____

_____

Astronomer's Journal with 100 observational logs

Blue nebula edition, 8" x 10"

Copyright © 2018 Camp Galaxy

ISBN: 978-1-73-128188-3

Astronomer's Journal

Welcome to your astronomer's journal! Taking notes on astronomical observations is a valuable tool and time-honored tradition for professional and amateur astronomers alike. We hope that this journal serves you well in recording your astronomical observations and sharing your experiences with others.

This journal includes 100 blank observational logs to enhance your skygazing experience. Designed by Camp Galaxy, the elegant log format included here allows you to record details about the object or objects being observed and the site and conditions at the time of observation. Five full pages for notes and reflections are also included at the end of the journal.

Paperback-bound with a charming and beautiful cover, this journal is sure to be a lasting record of your unique observations.

Are we human because we gaze at the stars,
or do we gaze at them because we are human?

–Neil Gaiman, *Stardust*

Observation Log

Object: _____

Constellation: _____

Instrument: _____

_____

Aperture: _____

Focal Length: _____

Eyepiece/Magnification:

_____

Filter(s): _____

Finder

Observer: _____

Date: _____

Time: _____

Location: _____

Seeing: _____

Transparency: _____

Lunar Phase: _____

Conditions: _____

_____

EP: _____ Mag: _____

Filter: _____ FOV: _____

EP: _____ Mag: _____

Filter: _____ FOV: _____

Notes

_____
_____
_____
_____
_____
_____
_____
_____
_____
_____
_____
_____

Observation Log

Object: _____

Constellation: _____

Instrument: _____

_____

Aperture: _____

Focal Length: _____

Eyepiece/Magnification:

_____

Filter(s): _____

Finder

Observer: _____

Date: _____

Time: _____

Location: _____

Seeing: _____

Transparency: _____

Lunar Phase: _____

Conditions: _____

_____

EP: _____   Mag: _____

Filter: _____   FOV: _____

EP: _____   Mag: _____

Filter: _____   FOV: _____

Notes

Observation Log

Object: _____

Constellation: _____

Instrument: _____

_____

Aperture: _____

Focal Length: _____

Eyepiece/Magnification:

_____

Filter(s): _____

Finder

Observer: _____

Date: _____

Time: _____

Location: _____

Seeing: _____

Transparency: _____

Lunar Phase: _____

Conditions: _____

_____

EP: _____ Mag: _____

Filter: _____ FOV: _____

EP: _____ Mag: _____

Filter: _____ FOV: _____

Notes

_____

_____

_____

_____

_____

_____

_____

_____

_____

_____

_____

_____

Observation Log

Object: _____

Constellation: _____

Instrument: _____

_____

Aperture: _____

Focal Length: _____

Eyepiece/Magnification:

_____

Filter(s): _____

Finder

Observer: _____

Date: _____

Time: _____

Location: _____

Seeing: _____

Transparency: _____

Lunar Phase: _____

Conditions: _____

_____

EP: _____ Mag: _____   EP: _____ Mag: _____

Filter: _____ FOV: _____   Filter: _____ FOV: _____

Notes

_____

_____

_____

_____

_____

_____

_____

_____

_____

_____

_____

_____

_____

_____

Observation Log

Object: _____

Constellation: _____

Instrument: _____

_____

Aperture: _____

Focal Length: _____

Eyepiece/Magnification:

_____

Filter(s): _____

Finder

Observer: _____

Date: _____

Time: _____

Location: _____

Seeing: _____

Transparency: _____

Lunar Phase: _____

Conditions: _____

_____

EP: _____ Mag: _____

Filter: _____ FOV: _____

EP: _____ Mag: _____

Filter: _____ FOV: _____

Notes

_____

_____

_____

_____

_____

_____

_____

_____

_____

_____

_____

_____

_____

Observation Log

Object: _____

Constellation: _____

Instrument: _____

_____

Aperture: _____

Focal Length: _____

Eyepiece/Magnification:

_____

Filter(s): _____

Finder

Observer: _____

Date: _____

Time: _____

Location: _____

Seeing: _____

Transparency: _____

Lunar Phase: _____

Conditions: _____

_____

EP: _____ Mag: _____

Filter: _____ FOV: _____

EP: _____ Mag: _____

Filter: _____ FOV: _____

Notes

Observation Log

Object: _____

Constellation: _____

Instrument: _____

_____

Aperture: _____

Focal Length: _____

Eyepiece/Magnification:

_____

Filter(s): _____

Finder

Observer: _____

Date: _____

Time: _____

Location: _____

Seeing: _____

Transparency: _____

Lunar Phase: _____

Conditions: _____

_____

EP: _____   Mag: _____

Filter: _____   FOV: _____

EP: _____   Mag: _____

Filter: _____   FOV: _____

Notes

_____

_____

_____

_____

_____

_____

_____

_____

_____

_____

_____

_____

Observation Log

Object: _____

Constellation: _____

Instrument: _____

_____

Aperture: _____

Focal Length: _____

Eyepiece/Magnification:

_____

Filter(s): _____

Finder

Observer: _____

Date: _____

Time: _____

Location: _____

Seeing: _____

Transparency: _____

Lunar Phase: _____

Conditions: _____

_____

EP: _____ Mag: _____

Filter: _____ FOV: _____

EP: _____ Mag: _____

Filter: _____ FOV: _____

Notes

Observation Log

Object: _____

Constellation: _____

Instrument: _____

_____

Aperture: _____

Focal Length: _____

Eyepiece/Magnification:

_____

Filter(s): _____

Finder

Observer: _____

Date: _____

Time: _____

Location: _____

Seeing: _____

Transparency: _____

Lunar Phase: _____

Conditions: _____

_____

EP: _____   Mag: _____

Filter: _____   FOV: _____

EP: _____   Mag: _____

Filter: _____   FOV: _____

Notes

_____

_____

_____

_____

_____

_____

_____

_____

_____

_____

_____

_____

Observation Log

Object: _____

Constellation: _____

Instrument: _____

_____

Aperture: _____

Focal Length: _____

Eyepiece/Magnification:

_____

Filter(s): _____

Finder

Observer: _____

Date: _____

Time: _____

Location: _____

Seeing: _____

Transparency: _____

Lunar Phase: _____

Conditions: _____

_____

EP: _____   Mag: _____

Filter: _____   FOV: _____

EP: _____   Mag: _____

Filter: _____   FOV: _____

Notes

_____

_____

_____

_____

_____

_____

_____

_____

_____

_____

_____

_____

_____

Observation Log

Object: _____

Constellation: _____

Instrument: _____

_____

Aperture: _____

Focal Length: _____

Eyepiece/Magnification:

_____

Filter(s): _____

Finder

Observer: _____

Date: _____

Time: _____

Location: _____

Seeing: _____

Transparency: _____

Lunar Phase: _____

Conditions: _____

_____

EP: _____   Mag: _____

Filter: _____   FOV: _____

EP: _____   Mag: _____

Filter: _____   FOV: _____

Notes

Observation Log

Object: _____

Constellation: _____

Instrument: _____

_____

Aperture: _____

Focal Length: _____

Eyepiece/Magnification:

_____

Filter(s): _____

Finder

Observer: _____

Date: _____

Time: _____

Location: _____

Seeing: _____

Transparency: _____

Lunar Phase: _____

Conditions: _____

_____

EP: _____   Mag: _____

Filter: _____   FOV: _____

EP: _____   Mag: _____

Filter: _____   FOV: _____

Notes

_____

_____

_____

_____

_____

_____

_____

_____

_____

_____

_____

_____

Observation Log

Object: _____

Constellation: _____

Instrument: _____

_____

Aperture: _____

Focal Length: _____

Eyepiece/Magnification:

_____

Filter(s): _____

Finder

Observer: _____

Date: _____

Time: _____

Location: _____

Seeing: _____

Transparency: _____

Lunar Phase: _____

Conditions: _____

_____

EP: _____ Mag: _____

Filter: _____ FOV: _____

EP: _____ Mag: _____

Filter: _____ FOV: _____

Notes

Observation Log

Object: _____

Constellation: _____

Instrument: _____

_____

Aperture: _____

Focal Length: _____

Eyepiece/Magnification:

_____

Filter(s): _____

Finder

Observer: _____

Date: _____

Time: _____

Location: _____

Seeing: _____

Transparency: _____

Lunar Phase: _____

Conditions: _____

_____

EP: _____ Mag: _____

Filter: _____ FOV: _____

EP: _____ Mag: _____

Filter: _____ FOV: _____

Notes

Observation Log

Object: _____

Constellation: _____

Instrument: _____

_____

Aperture: _____

Focal Length: _____

Eyepiece/Magnification: _____

_____

Filter(s): _____

Finder

Observer: _____

Date: _____

Time: _____

Location: _____

Seeing: _____

Transparency: _____

Lunar Phase: _____

Conditions: _____

_____

EP: _____ Mag: _____

Filter: _____ FOV: _____

EP: _____ Mag: _____

Filter: _____ FOV: _____

Notes

Observation Log

Object: _____

Constellation: _____

Instrument: _____

_____

Aperture: _____

Focal Length: _____

Eyepiece/Magnification:

_____

Filter(s): _____

Finder

Observer: _____

Date: _____

Time: _____

Location: _____

Seeing: _____

Transparency: _____

Lunar Phase: _____

Conditions: _____

_____

EP: _____   Mag: _____

Filter: _____   FOV: _____

EP: _____   Mag: _____

Filter: _____   FOV: _____

Notes

_____

_____

_____

_____

_____

_____

_____

_____

_____

_____

_____

Observation Log

Object: _____

Constellation: _____

Instrument: _____

_____

Aperture: _____

Focal Length: _____

Eyepiece/Magnification:

_____

Filter(s): _____

Finder

Observer: _____

Date: _____

Time: _____

Location: _____

Seeing: _____

Transparency: _____

Lunar Phase: _____

Conditions: _____

_____

EP: _____ Mag: _____   EP: _____ Mag: _____

Filter: _____ FOV: _____   Filter: _____ FOV: _____

Notes

Observation Log

Object: _____

Constellation: _____

Instrument: _____

_____

Aperture: _____

Focal Length: _____

Eyepiece/Magnification:

_____

Filter(s): _____

Finder

Observer: _____

Date: _____

Time: _____

Location: _____

Seeing: _____

Transparency: _____

Lunar Phase: _____

Conditions: _____

_____

EP: _____ Mag: _____

Filter: _____ FOV: _____

EP: _____ Mag: _____

Filter: _____ FOV: _____

Notes

_____

_____

_____

_____

_____

_____

_____

_____

_____

_____

_____

_____

Observation Log

Object: _____

Constellation: _____

Instrument: _____

_____

Aperture: _____

Focal Length: _____

Eyepiece/Magnification:

_____

Filter(s): _____

Finder

Observer: _____

Date: _____

Time: _____

Location: _____

Seeing: _____

Transparency: _____

Lunar Phase: _____

Conditions: _____

_____

EP: _____ Mag: _____

Filter: _____ FOV: _____

EP: _____ Mag: _____

Filter: _____ FOV: _____

Notes

Observation Log

Object: _____

Constellation: _____

Instrument: _____

_____

Aperture: _____

Focal Length: _____

Eyepiece/Magnification:

_____

Filter(s): _____

Finder

Observer: _____

Date: _____

Time: _____

Location: _____

Seeing: _____

Transparency: _____

Lunar Phase: _____

Conditions: _____

_____

EP: _____   Mag: _____

Filter: _____   FOV: _____

EP: _____   Mag: _____

Filter: _____   FOV: _____

Notes

Observation Log

Object: _____

Constellation: _____

Instrument: _____

_____

Aperture: _____

Focal Length: _____

Eyepiece/Magnification:

_____

Filter(s): _____

Finder

Observer: _____

Date: _____

Time: _____

Location: _____

Seeing: _____

Transparency: _____

Lunar Phase: _____

Conditions: _____

_____

EP: _____   Mag: _____

Filter: _____   FOV: _____

EP: _____   Mag: _____

Filter: _____   FOV: _____

Notes

_____

_____

_____

_____

_____

_____

_____

_____

_____

_____

_____

_____

Observation Log

Object: _____

Constellation: _____

Instrument: _____

_____

Aperture: _____

Focal Length: _____

Eyepiece/Magnification:

_____

Filter(s): _____

Finder

Observer: _____

Date: _____

Time: _____

Location: _____

Seeing: _____

Transparency: _____

Lunar Phase: _____

Conditions: _____

_____

EP: _____   Mag: _____

Filter: _____   FOV: _____

EP: _____   Mag: _____

Filter: _____   FOV: _____

Notes

_____

_____

_____

_____

_____

_____

_____

_____

_____

_____

_____

_____

Observation Log

Object: _____

Constellation: _____

Instrument: _____

_____

Aperture: _____

Focal Length: _____

Eyepiece/Magnification:

_____

Filter(s): _____

Finder

Observer: _____

Date: _____

Time: _____

Location: _____

Seeing: _____

Transparency: _____

Lunar Phase: _____

Conditions: _____

_____

EP: _____ Mag: _____

Filter: _____ FOV: _____

EP: _____ Mag: _____

Filter: _____ FOV: _____

Notes

Observation Log

Object: _____

Constellation: _____

Instrument: _____

_____

Aperture: _____

Focal Length: _____

Eyepiece/Magnification:

_____

Filter(s): _____

Finder

Observer: _____

Date: _____

Time: _____

Location: _____

Seeing: _____

Transparency: _____

Lunar Phase: _____

Conditions: _____

_____

EP: _____ Mag: _____

Filter: _____ FOV: _____

EP: _____ Mag: _____

Filter: _____ FOV: _____

Notes

_____

_____

_____

_____

_____

_____

_____

_____

_____

_____

_____

_____

Observation Log

Object: _____

Constellation: _____

Instrument: _____

_____

Aperture: _____

Focal Length: _____

Eyepiece/Magnification:

_____

Filter(s): _____

Finder

Observer: _____

Date: _____

Time: _____

Location: _____

Seeing: _____

Transparency: _____

Lunar Phase: _____

Conditions: _____

_____

EP: _____   Mag: _____

Filter: _____   FOV: _____

EP: _____   Mag: _____

Filter: _____   FOV: _____

Notes

Observation Log

Object: _____

Constellation: _____

Instrument: _____

_____

Aperture: _____

Focal Length: _____

Eyepiece/Magnification:

_____

Filter(s): _____

Finder

Observer: _____

Date: _____

Time: _____

Location: _____

Seeing: _____

Transparency: _____

Lunar Phase: _____

Conditions: _____

_____

EP: _____   Mag: _____

Filter: _____   FOV: _____

EP: _____   Mag: _____

Filter: _____   FOV: _____

Notes

_____

_____

_____

_____

_____

_____

_____

_____

_____

_____

_____

_____

Observation Log

Object: _____

Constellation: _____

Instrument: _____

_____

Aperture: _____

Focal Length: _____

Eyepiece/Magnification:

_____

Filter(s): _____

Finder

Observer: _____

Date: _____

Time: _____

Location: _____

Seeing: _____

Transparency: _____

Lunar Phase: _____

Conditions: _____

_____

EP: _____   Mag: _____

Filter: _____   FOV: _____

EP: _____   Mag: _____

Filter: _____   FOV: _____

Notes

_____

_____

_____

_____

_____

_____

_____

_____

_____

_____

_____

_____

Observation Log

Object: _____

Constellation: _____

Instrument: _____

_____

Aperture: _____

Focal Length: _____

Eyepiece/Magnification:

_____

Filter(s): _____

Finder

Observer: _____

Date: _____

Time: _____

Location: _____

Seeing: _____

Transparency: _____

Lunar Phase: _____

Conditions: _____

_____

EP: _____   Mag: _____

Filter: _____   FOV: _____

EP: _____   Mag: _____

Filter: _____   FOV: _____

Notes

_____

_____

_____

_____

_____

_____

_____

_____

_____

_____

_____

_____

_____

Observation Log

Object: _____

Constellation: _____

Instrument: _____

_____

Aperture: _____

Focal Length: _____

Eyepiece/Magnification:

_____

Filter(s): _____

Finder

Observer: _____

Date: _____

Time: _____

Location: _____

Seeing: _____

Transparency: _____

Lunar Phase: _____

Conditions: _____

_____

EP: _____ Mag: _____

Filter: _____ FOV: _____

EP: _____ Mag: _____

Filter: _____ FOV: _____

Notes

_____
_____
_____
_____
_____
_____
_____
_____
_____
_____
_____
_____
_____

Observation Log

Object: _____

Constellation: _____

Instrument: _____

_____

Aperture: _____

Focal Length: _____

Eyepiece/Magnification:

_____

Filter(s): _____

Finder

Observer: _____

Date: _____

Time: _____

Location: _____

Seeing: _____

Transparency: _____

Lunar Phase: _____

Conditions: _____

_____

EP: _____ Mag: _____

Filter: _____ FOV: _____

EP: _____ Mag: _____

Filter: _____ FOV: _____

Notes

_____
_____
_____
_____
_____
_____
_____
_____
_____
_____
_____
_____
_____

Observation Log

Object: _____

Constellation: _____

Instrument: _____

_____

Aperture: _____

Focal Length: _____

Eyepiece/Magnification:

_____

Filter(s): _____

Finder

Observer: _____

Date: _____

Time: _____

Location: _____

Seeing: _____

Transparency: _____

Lunar Phase: _____

Conditions: _____

_____

EP: _____   Mag: _____   EP: _____   Mag: _____

Filter: _____   FOV: _____   Filter: _____   FOV: _____

Notes

_____

_____

_____

_____

_____

_____

_____

_____

_____

_____

_____

_____

Observation Log

Object: _____

Constellation: _____

Instrument: _____

_____

Aperture: _____

Focal Length: _____

Eyepiece/Magnification:

_____

Filter(s): _____

Finder

Observer: _____

Date: _____

Time: _____

Location: _____

Seeing: _____

Transparency: _____

Lunar Phase: _____

Conditions: _____

_____

EP: _____ Mag: _____

Filter: _____ FOV: _____

EP: _____ Mag: _____

Filter: _____ FOV: _____

Notes

Observation Log

Object: _____

Constellation: _____

Instrument: _____

_____

Aperture: _____

Focal Length: _____

Eyepiece/Magnification:

_____

Filter(s): _____

Finder

Observer: _____

Date: _____

Time: _____

Location: _____

Seeing: _____

Transparency: _____

Lunar Phase: _____

Conditions: _____

_____

EP: _____   Mag: _____

Filter: _____   FOV: _____

EP: _____   Mag: _____

Filter: _____   FOV: _____

Notes

Observation Log

Object: _____

Constellation: _____

Instrument: _____

_____

Aperture: _____

Focal Length: _____

Eyepiece/Magnification:

_____

Filter(s): _____

Finder

Observer: _____

Date: _____

Time: _____

Location: _____

Seeing: _____

Transparency: _____

Lunar Phase: _____

Conditions: _____

_____

EP: _____ Mag: _____

Filter: _____ FOV: _____

EP: _____ Mag: _____

Filter: _____ FOV: _____

Notes

Observation Log

Object: _____

Constellation: _____

Instrument: _____

_____

Aperture: _____

Focal Length: _____

Eyepiece/Magnification:

_____

Filter(s): _____

Finder

Observer: _____

Date: _____

Time: _____

Location: _____

Seeing: _____

Transparency: _____

Lunar Phase: _____

Conditions: _____

_____

EP: _____   Mag: _____

Filter: _____   FOV: _____

EP: _____   Mag: _____

Filter: _____   FOV: _____

Notes

Observation Log

Object: _____

Constellation: _____

Instrument: _____

_____

Aperture: _____

Focal Length: _____

Eyepiece/Magnification:

_____

Filter(s): _____

Finder

Observer: _____

Date: _____

Time: _____

Location: _____

Seeing: _____

Transparency: _____

Lunar Phase: _____

Conditions: _____

_____

EP: _____ Mag: _____

Filter: _____ FOV: _____

EP: _____ Mag: _____

Filter: _____ FOV: _____

Notes

_____

_____

_____

_____

_____

_____

_____

_____

_____

_____

_____

_____

_____

_____

Observation Log

Object: _____

Constellation: _____

Instrument: _____

_____

Aperture: _____

Focal Length: _____

Eyepiece/Magnification:

_____

Filter(s): _____

Finder

Observer: _____

Date: _____

Time: _____

Location: _____

Seeing: _____

Transparency: _____

Lunar Phase: _____

Conditions: _____

_____

EP: _____   Mag: _____

Filter: _____   FOV: _____

EP: _____   Mag: _____

Filter: _____   FOV: _____

Notes

Observation Log

Object: _____

Constellation: _____

Instrument: _____

_____

Aperture: _____

Focal Length: _____

Eyepiece/Magnification:

_____

Filter(s): _____

Observer: _____

Date: _____

Time: _____

Location: _____

Seeing: _____

Transparency: _____

Lunar Phase: _____

Conditions: _____

_____

Finder

EP: _____   Mag: _____

Filter: _____   FOV: _____

EP: _____   Mag: _____

Filter: _____   FOV: _____

Notes

Observation Log

Object: _____

Constellation: _____

Instrument: _____

_____

Aperture: _____

Focal Length: _____

Eyepiece/Magnification:

_____

Filter(s): _____

Finder

Observer: _____

Date: _____

Time: _____

Location: _____

Seeing: _____

Transparency: _____

Lunar Phase: _____

Conditions: _____

_____

EP: _____ Mag: _____

Filter: _____ FOV: _____

EP: _____ Mag: _____

Filter: _____ FOV: _____

Notes

Observation Log

Object: _____

Constellation: _____

Instrument: _____

_____

Aperture: _____

Focal Length: _____

Eyepiece/Magnification:

_____

Filter(s): _____

Finder

Observer: _____

Date: _____

Time: _____

Location: _____

Seeing: _____

Transparency: _____

Lunar Phase: _____

Conditions: _____

_____

EP: _____   Mag: _____

Filter: _____   FOV: _____

EP: _____   Mag: _____

Filter: _____   FOV: _____

Notes

Observation Log

Object: _____

Constellation: _____

Instrument: _____

_____

Aperture: _____

Focal Length: _____

Eyepiece/Magnification:

_____

Filter(s): _____

Finder

Observer: _____

Date: _____

Time: _____

Location: _____

Seeing: _____

Transparency: _____

Lunar Phase: _____

Conditions: _____

_____

EP: _____   Mag: _____

Filter: _____   FOV: _____

EP: _____   Mag: _____

Filter: _____   FOV: _____

Notes

_____
_____
_____
_____
_____
_____
_____
_____
_____
_____
_____
_____

Observation Log

Object: _____

Constellation: _____

Instrument: _____

_____

Aperture: _____

Focal Length: _____

Eyepiece/Magnification:

_____

Filter(s): _____

Finder

Observer: _____

Date: _____

Time: _____

Location: _____

Seeing: _____

Transparency: _____

Lunar Phase: _____

Conditions: _____

_____

EP: _____   Mag: _____

Filter: _____   FOV: _____

EP: _____   Mag: _____

Filter: _____   FOV: _____

Notes

Observation Log

Object: _____

Constellation: _____

Instrument: _____

_____

Aperture: _____

Focal Length: _____

Eyepiece/Magnification: _____

_____

Filter(s): _____

Finder

Observer: _____

Date: _____

Time: _____

Location: _____

Seeing: _____

Transparency: _____

Lunar Phase: _____

Conditions: _____

_____

EP: _____   Mag: _____

Filter: _____   FOV: _____

EP: _____   Mag: _____

Filter: _____   FOV: _____

Notes

_____

_____

_____

_____

_____

_____

_____

_____

_____

_____

_____

Observation Log

Object: _____

Constellation: _____

Instrument: _____

_____

Aperture: _____

Focal Length: _____

Eyepiece/Magnification:

_____

Filter(s): _____

Finder

Observer: _____

Date: _____

Time: _____

Location: _____

Seeing: _____

Transparency: _____

Lunar Phase: _____

Conditions: _____

_____

EP: _____   Mag: _____

Filter: _____   FOV: _____

EP: _____   Mag: _____

Filter: _____   FOV: _____

Notes

_____

_____

_____

_____

_____

_____

_____

_____

_____

_____

_____

_____

Observation Log

Object: _____

Constellation: _____

Instrument: _____

_____

Aperture: _____

Focal Length: _____

Eyepiece/Magnification: _____

_____

Filter(s): _____

Finder

Observer: _____

Date: _____

Time: _____

Location: _____

Seeing: _____

Transparency: _____

Lunar Phase: _____

Conditions: _____

_____

EP: _____   Mag: _____

Filter: _____   FOV: _____

EP: _____   Mag: _____

Filter: _____   FOV: _____

Notes

_____

_____

_____

_____

_____

_____

_____

_____

_____

_____

_____

_____

Observation Log

Object: _____

Constellation: _____

Instrument: _____

_____

Aperture: _____

Focal Length: _____

Eyepiece/Magnification:

_____

Filter(s): _____

Finder

Observer: _____

Date: _____

Time: _____

Location: _____

Seeing: _____

Transparency: _____

Lunar Phase: _____

Conditions: _____

_____

EP: _____   Mag: _____

Filter: _____   FOV: _____

EP: _____   Mag: _____

Filter: _____   FOV: _____

Notes

Observation Log

Object: _____

Constellation: _____

Instrument: _____

_____

Aperture: _____

Focal Length: _____

Eyepiece/Magnification:

_____

Filter(s): _____

Finder

Observer: _____

Date: _____

Time: _____

Location: _____

Seeing: _____

Transparency: _____

Lunar Phase: _____

Conditions: _____

_____

EP: _____   Mag: _____

Filter: _____   FOV: _____

EP: _____   Mag: _____

Filter: _____   FOV: _____

Notes

Observation Log

Object: _____

Constellation: _____

Instrument: _____

_____

Aperture: _____

Focal Length: _____

Eyepiece/Magnification:

_____

Filter(s): _____

Finder

Observer: _____

Date: _____

Time: _____

Location: _____

Seeing: _____

Transparency: _____

Lunar Phase: _____

Conditions: _____

_____

EP: _____ Mag: _____

Filter: _____ FOV: _____

EP: _____ Mag: _____

Filter: _____ FOV: _____

Notes

_____
_____
_____
_____
_____
_____
_____
_____
_____
_____
_____
_____
_____
_____

Observation Log

Object: _____

Constellation: _____

Instrument: _____

_____

Aperture: _____

Focal Length: _____

Eyepiece/Magnification:

_____

Filter(s): _____

Finder

Observer: _____

Date: _____

Time: _____

Location: _____

Seeing: _____

Transparency: _____

Lunar Phase: _____

Conditions: _____

_____

EP: _____ Mag: _____

Filter: _____ FOV: _____

EP: _____ Mag: _____

Filter: _____ FOV: _____

Notes

Observation Log

Object: _____

Constellation: _____

Instrument: _____

_____

Aperture: _____

Focal Length: _____

Eyepiece/Magnification:

_____

Filter(s): _____

Finder

Observer: _____

Date: _____

Time: _____

Location: _____

Seeing: _____

Transparency: _____

Lunar Phase: _____

Conditions: _____

_____

EP: _____ Mag: _____

Filter: _____ FOV: _____

EP: _____ Mag: _____

Filter: _____ FOV: _____

Notes

_____
_____
_____
_____
_____
_____
_____
_____
_____
_____
_____
_____
_____
_____

Observation Log

Object: _____

Constellation: _____

Instrument: _____

_____

Aperture: _____

Focal Length: _____

Eyepiece/Magnification: _____

_____

Filter(s): _____

Finder

Observer: _____

Date: _____

Time: _____

Location: _____

Seeing: _____

Transparency: _____

Lunar Phase: _____

Conditions: _____

_____

EP: _____   Mag: _____

Filter: _____   FOV: _____

EP: _____   Mag: _____

Filter: _____   FOV: _____

Notes

_____
_____
_____
_____
_____
_____
_____
_____
_____
_____
_____
_____

Observation Log

Object: _____

Constellation: _____

Instrument: _____

_____

Aperture: _____

Focal Length: _____

Eyepiece/Magnification:

_____

Filter(s): _____

Finder

Observer: _____

Date: _____

Time: _____

Location: _____

Seeing: _____

Transparency: _____

Lunar Phase: _____

Conditions: _____

_____

EP: _____ Mag: _____

Filter: _____ FOV: _____

EP: _____ Mag: _____

Filter: _____ FOV: _____

Notes

_____

_____

_____

_____

_____

_____

_____

_____

_____

_____

_____

_____

_____

Observation Log

Object: _____

Constellation: _____

Instrument: _____

_____

Aperture: _____

Focal Length: _____

Eyepiece/Magnification:

_____

Filter(s): _____

Finder

Observer: _____

Date: _____

Time: _____

Location: _____

Seeing: _____

Transparency: _____

Lunar Phase: _____

Conditions: _____

_____

EP: _____ Mag: _____

Filter: _____ FOV: _____

EP: _____ Mag: _____

Filter: _____ FOV: _____

Notes

_____

_____

_____

_____

_____

_____

_____

_____

_____

_____

_____

_____

_____

Observation Log

Object: _____

Constellation: _____

Instrument: _____

_____

Aperture: _____

Focal Length: _____

Eyepiece/Magnification:

_____

Filter(s): _____

Observer: _____

Date: _____

Time: _____

Location: _____

Seeing: _____

Transparency: _____

Lunar Phase: _____

Conditions: _____

_____

Finder

EP: _____   Mag: _____

Filter: _____   FOV: _____

EP: _____   Mag: _____

Filter: _____   FOV: _____

Notes

_____

_____

_____

_____

_____

_____

_____

_____

_____

_____

_____

Observation Log

Object: _____

Constellation: _____

Instrument: _____

_____

Aperture: _____

Focal Length: _____

Eyepiece/Magnification:

_____

Filter(s): _____

Finder

Observer: _____

Date: _____

Time: _____

Location: _____

Seeing: _____

Transparency: _____

Lunar Phase: _____

Conditions: _____

_____

EP: _____ Mag: _____

Filter: _____ FOV: _____

EP: _____ Mag: _____

Filter: _____ FOV: _____

Notes

_____
_____
_____
_____
_____
_____
_____
_____
_____
_____
_____
_____

Observation Log

Object: _____

Constellation: _____

Instrument: _____

_____

Aperture: _____

Focal Length: _____

Eyepiece/Magnification:

_____

Filter(s): _____

Finder

Observer: _____

Date: _____

Time: _____

Location: _____

Seeing: _____

Transparency: _____

Lunar Phase: _____

Conditions: _____

_____

EP: _____   Mag: _____

Filter: _____   FOV: _____

EP: _____   Mag: _____

Filter: _____   FOV: _____

Notes

_____

_____

_____

_____

_____

_____

_____

_____

_____

_____

_____

_____

Observation Log

Object: _____

Constellation: _____

Instrument: _____

_____

Aperture: _____

Focal Length: _____

Eyepiece/Magnification:

_____

Filter(s): _____

Finder

Observer: _____

Date: _____

Time: _____

Location: _____

Seeing: _____

Transparency: _____

Lunar Phase: _____

Conditions: _____

_____

EP: _____ Mag: _____

Filter: _____ FOV: _____

EP: _____ Mag: _____

Filter: _____ FOV: _____

Notes

_____

_____

_____

_____

_____

_____

_____

_____

_____

_____

_____

_____

Observation Log

Object: _____

Constellation: _____

Instrument: _____

_____

Aperture: _____

Focal Length: _____

Eyepiece/Magnification:

_____

Filter(s): _____

Finder

Observer: _____

Date: _____

Time: _____

Location: _____

Seeing: _____

Transparency: _____

Lunar Phase: _____

Conditions: _____

_____

EP: _____ Mag: _____

Filter: _____ FOV: _____

EP: _____ Mag: _____

Filter: _____ FOV: _____

Notes

Observation Log

Object: _____

Constellation: _____

Instrument: _____

_____

Aperture: _____

Focal Length: _____

Eyepiece/Magnification: _____

_____

Filter(s): _____

Finder

Observer: _____

Date: _____

Time: _____

Location: _____

Seeing: _____

Transparency: _____

Lunar Phase: _____

Conditions: _____

_____

EP: _____ Mag: _____

Filter: _____ FOV: _____

EP: _____ Mag: _____

Filter: _____ FOV: _____

Notes

_____
_____
_____
_____
_____
_____
_____
_____
_____
_____
_____
_____

Observation Log

Object: _____

Constellation: _____

Instrument: _____

_____

Aperture: _____

Focal Length: _____

Eyepiece/Magnification:

_____

Filter(s): _____

Finder

Observer: _____

Date: _____

Time: _____

Location: _____

Seeing: _____

Transparency: _____

Lunar Phase: _____

Conditions: _____

_____

EP: _____   Mag: _____

Filter: _____   FOV: _____

EP: _____   Mag: _____

Filter: _____   FOV: _____

Notes

Observation Log

Object: _____

Constellation: _____

Instrument: _____

_____

Aperture: _____

Focal Length: _____

Eyepiece/Magnification:

_____

Filter(s): _____

Finder

Observer: _____

Date: _____

Time: _____

Location: _____

Seeing: _____

Transparency: _____

Lunar Phase: _____

Conditions: _____

_____

EP: _____ Mag: _____

Filter: _____ FOV: _____

EP: _____ Mag: _____

Filter: _____ FOV: _____

Notes

_____

_____

_____

_____

_____

_____

_____

_____

_____

_____

_____

_____

Observation Log

Object: _____

Constellation: _____

Instrument: _____

_____

Aperture: _____

Focal Length: _____

Eyepiece/Magnification:

_____

Filter(s): _____

Finder

Observer: _____

Date: _____

Time: _____

Location: _____

Seeing: _____

Transparency: _____

Lunar Phase: _____

Conditions: _____

_____

EP: _____ Mag: _____

Filter: _____ FOV: _____

EP: _____ Mag: _____

Filter: _____ FOV: _____

Notes

_____

_____

_____

_____

_____

_____

_____

_____

_____

_____

_____

_____

_____

Observation Log

Object: _____

Constellation: _____

Instrument: _____

_____

Aperture: _____

Focal Length: _____

Eyepiece/Magnification: _____

_____

Filter(s): _____

Finder

Observer: _____

Date: _____

Time: _____

Location: _____

Seeing: _____

Transparency: _____

Lunar Phase: _____

Conditions: _____

_____

EP: _____ Mag: _____

Filter: _____ FOV: _____

EP: _____ Mag: _____

Filter: _____ FOV: _____

Notes

_____

_____

_____

_____

_____

_____

_____

_____

_____

_____

_____

_____

Observation Log

Object: _____

Constellation: _____

Instrument: _____

_____

Aperture: _____

Focal Length: _____

Eyepiece/Magnification:

_____

Filter(s): _____

Finder

Observer: _____

Date: _____

Time: _____

Location: _____

Seeing: _____

Transparency: _____

Lunar Phase: _____

Conditions: _____

_____

EP: _____   Mag: _____

Filter: _____   FOV: _____

EP: _____   Mag: _____

Filter: _____   FOV: _____

Notes

_____
_____
_____
_____
_____
_____
_____
_____
_____
_____
_____
_____
_____
_____

Observation Log

Object: _____

Constellation: _____

Instrument: _____

_____

Aperture: _____

Focal Length: _____

Eyepiece/Magnification:

_____

Filter(s): _____

Finder

Observer: _____

Date: _____

Time: _____

Location: _____

Seeing: _____

Transparency: _____

Lunar Phase: _____

Conditions: _____

_____

EP: _____   Mag: _____

Filter: _____   FOV: _____

EP: _____   Mag: _____

Filter: _____   FOV: _____

Notes

Observation Log

Object: _____

Constellation: _____

Instrument: _____

_____

Aperture: _____

Focal Length: _____

Eyepiece/Magnification:

_____

Filter(s): _____

Finder

Observer: _____

Date: _____

Time: _____

Location: _____

Seeing: _____

Transparency: _____

Lunar Phase: _____

Conditions: _____

_____

EP: _____   Mag: _____

Filter: _____   FOV: _____

EP: _____   Mag: _____

Filter: _____   FOV: _____

Notes

Observation Log

Object: _____

Constellation: _____

Instrument: _____

_____

Aperture: _____

Focal Length: _____

Eyepiece/Magnification:

_____

Filter(s): _____

Observer: _____

Date: _____

Time: _____

Location: _____

Seeing: _____

Transparency: _____

Lunar Phase: _____

Conditions: _____

_____

Finder

EP: _____   Mag: _____

Filter: _____   FOV: _____

EP: _____   Mag: _____

Filter: _____   FOV: _____

Notes

Observation Log

Object: _____

Constellation: _____

Instrument: _____

_____

Aperture: _____

Focal Length: _____

Eyepiece/Magnification:

_____

Filter(s): _____

Finder

Observer: _____

Date: _____

Time: _____

Location: _____

Seeing: _____

Transparency: _____

Lunar Phase: _____

Conditions: _____

_____

EP: _____ Mag: _____

Filter: _____ FOV: _____

EP: _____ Mag: _____

Filter: _____ FOV: _____

Notes

Observation Log

Object: _____

Constellation: _____

Instrument: _____

_____

Aperture: _____

Focal Length: _____

Eyepiece/Magnification:

_____

Filter(s): _____

Finder

Observer: _____

Date: _____

Time: _____

Location: _____

Seeing: _____

Transparency: _____

Lunar Phase: _____

Conditions: _____

_____

EP: _____ Mag: _____

Filter: _____ FOV: _____

EP: _____ Mag: _____

Filter: _____ FOV: _____

Notes

_____

_____

_____

_____

_____

_____

_____

_____

_____

_____

_____

_____

Observation Log

Object: _____

Constellation: _____

Instrument: _____

_____

Aperture: _____

Focal Length: _____

Eyepiece/Magnification:

_____

Filter(s): _____

Finder

Observer: _____

Date: _____

Time: _____

Location: _____

Seeing: _____

Transparency: _____

Lunar Phase: _____

Conditions: _____

_____

EP: _____   Mag: _____

Filter: _____   FOV: _____

EP: _____   Mag: _____

Filter: _____   FOV: _____

Notes

_____
_____
_____
_____
_____
_____
_____
_____
_____
_____
_____
_____
_____

Observation Log

Object: _____

Constellation: _____

Instrument: _____

_____

Aperture: _____

Focal Length: _____

Eyepiece/Magnification: _____

_____

Filter(s): _____

Finder

Observer: _____

Date: _____

Time: _____

Location: _____

Seeing: _____

Transparency: _____

Lunar Phase: _____

Conditions: _____

_____

EP: _____ Mag: _____

Filter: _____ FOV: _____

EP: _____ Mag: _____

Filter: _____ FOV: _____

Notes

Observation Log

Object: _____

Constellation: _____

Instrument: _____

_____

Aperture: _____

Focal Length: _____

Eyepiece/Magnification:

_____

Filter(s): _____

Finder

Observer: _____

Date: _____

Time: _____

Location: _____

Seeing: _____

Transparency: _____

Lunar Phase: _____

Conditions: _____

_____

EP: _____   Mag: _____

Filter: _____   FOV: _____

EP: _____   Mag: _____

Filter: _____   FOV: _____

Notes

_____
_____
_____
_____
_____
_____
_____
_____
_____
_____
_____
_____

Observation Log

Object: _____

Constellation: _____

Instrument: _____

_____

Aperture: _____

Focal Length: _____

Eyepiece/Magnification: _____

_____

Filter(s): _____

Observer: _____

Date: _____

Time: _____

Location: _____

Seeing: _____

Transparency: _____

Lunar Phase: _____

Conditions: _____

_____

Finder

EP: _____   Mag: _____

Filter: _____   FOV: _____

EP: _____   Mag: _____

Filter: _____   FOV: _____

Notes

_____

_____

_____

_____

_____

_____

_____

_____

_____

_____

_____

_____

Observation Log

Object: _____

Constellation: _____

Instrument: _____

_____

Aperture: _____

Focal Length: _____

Eyepiece/Magnification:

_____

Filter(s): _____

Finder

Observer: _____

Date: _____

Time: _____

Location: _____

Seeing: _____

Transparency: _____

Lunar Phase: _____

Conditions: _____

_____

EP: _____ Mag: _____

Filter: _____ FOV: _____

EP: _____ Mag: _____

Filter: _____ FOV: _____

Notes

_____
_____
_____
_____
_____
_____
_____
_____
_____
_____
_____
_____
_____

Observation Log

Object: _____

Constellation: _____

Instrument: _____

_____

Aperture: _____

Focal Length: _____

Eyepiece/Magnification: _____

_____

Filter(s): _____

Finder

Observer: _____

Date: _____

Time: _____

Location: _____

Seeing: _____

Transparency: _____

Lunar Phase: _____

Conditions: _____

_____

EP: _____ Mag: _____

Filter: _____ FOV: _____

EP: _____ Mag: _____

Filter: _____ FOV: _____

Notes

_____

_____

_____

_____

_____

_____

_____

_____

_____

_____

_____

_____

Observation Log

Object: _____

Constellation: _____

Instrument: _____

_____

Aperture: _____

Focal Length: _____

Eyepiece/Magnification:

_____

Filter(s): _____

Finder

Observer: _____

Date: _____

Time: _____

Location: _____

Seeing: _____

Transparency: _____

Lunar Phase: _____

Conditions: _____

_____

EP: _____   Mag: _____

Filter: _____   FOV: _____

EP: _____   Mag: _____

Filter: _____   FOV: _____

Notes

_____

_____

_____

_____

_____

_____

_____

_____

_____

_____

_____

_____

_____

Observation Log

Object: _____

Constellation: _____

Instrument: _____

_____

Aperture: _____

Focal Length: _____

Eyepiece/Magnification:

_____

Filter(s): _____

Finder

Observer: _____

Date: _____

Time: _____

Location: _____

Seeing: _____

Transparency: _____

Lunar Phase: _____

Conditions: _____

_____

EP: _____   Mag: _____

Filter: _____   FOV: _____

EP: _____   Mag: _____

Filter: _____   FOV: _____

Notes

Observation Log

Object: _____

Constellation: _____

Instrument: _____

_____

Aperture: _____

Focal Length: _____

Eyepiece/Magnification:

_____

Filter(s): _____

Finder

Observer: _____

Date: _____

Time: _____

Location: _____

Seeing: _____

Transparency: _____

Lunar Phase: _____

Conditions: _____

_____

EP: _____   Mag: _____

Filter: _____   FOV: _____

EP: _____   Mag: _____

Filter: _____   FOV: _____

Notes

_____

_____

_____

_____

_____

_____

_____

_____

_____

_____

_____

_____

_____

Observation Log

Object: _____

Constellation: _____

Instrument: _____

_____

Aperture: _____

Focal Length: _____

Eyepiece/Magnification:

_____

Filter(s): _____

Finder

Observer: _____

Date: _____

Time: _____

Location: _____

Seeing: _____

Transparency: _____

Lunar Phase: _____

Conditions: _____

_____

EP: _____   Mag: _____

Filter: _____   FOV: _____

EP: _____   Mag: _____

Filter: _____   FOV: _____

Notes

_____

_____

_____

_____

_____

_____

_____

_____

_____

_____

_____

_____

Observation Log

Object: _____

Constellation: _____

Instrument: _____

_____

Aperture: _____

Focal Length: _____

Eyepiece/Magnification:

_____

Filter(s): _____

Finder

Observer: _____

Date: _____

Time: _____

Location: _____

Seeing: _____

Transparency: _____

Lunar Phase: _____

Conditions: _____

_____

EP: _____ Mag: _____

Filter: _____ FOV: _____

EP: _____ Mag: _____

Filter: _____ FOV: _____

Notes

_____

_____

_____

_____

_____

_____

_____

_____

_____

_____

_____

_____

Observation Log

Object: _____

Constellation: _____

Instrument: _____

_____

Aperture: _____

Focal Length: _____

Eyepiece/Magnification: _____

_____

Filter(s): _____

Finder

Observer: _____

Date: _____

Time: _____

Location: _____

Seeing: _____

Transparency: _____

Lunar Phase: _____

Conditions: _____

_____

EP: _____   Mag: _____

Filter: _____   FOV: _____

EP: _____   Mag: _____

Filter: _____   FOV: _____

Notes

_____

_____

_____

_____

_____

_____

_____

_____

_____

_____

Observation Log

Object: _____

Constellation: _____

Instrument: _____

_____

Aperture: _____

Focal Length: _____

Eyepiece/Magnification:

_____

Filter(s): _____

Finder

Observer: _____

Date: _____

Time: _____

Location: _____

Seeing: _____

Transparency: _____

Lunar Phase: _____

Conditions: _____

_____

EP: _____   Mag: _____

Filter: _____   FOV: _____

EP: _____   Mag: _____

Filter: _____   FOV: _____

Notes

_____

_____

_____

_____

_____

_____

_____

_____

_____

_____

_____

_____

Observation Log

Object: _____

Constellation: _____

Instrument: _____

_____

Aperture: _____

Focal Length: _____

Eyepiece/Magnification:

_____

Filter(s): _____

Finder

Observer: _____

Date: _____

Time: _____

Location: _____

Seeing: _____

Transparency: _____

Lunar Phase: _____

Conditions: _____

_____

EP: _____   Mag: _____

Filter: _____   FOV: _____

EP: _____   Mag: _____

Filter: _____   FOV: _____

Notes

_____
_____
_____
_____
_____
_____
_____
_____
_____
_____
_____
_____
_____

Observation Log

Object: _____

Constellation: _____

Instrument: _____

_____

Aperture: _____

Focal Length: _____

Eyepiece/Magnification:

_____

Filter(s): _____

Finder

Observer: _____

Date: _____

Time: _____

Location: _____

Seeing: _____

Transparency: _____

Lunar Phase: _____

Conditions: _____

_____

EP: _____   Mag: _____

Filter: _____   FOV: _____

EP: _____   Mag: _____

Filter: _____   FOV: _____

Notes

_____
_____
_____
_____
_____
_____
_____
_____
_____
_____
_____
_____

Observation Log

Object: _____

Constellation: _____

Instrument: _____

_____

Aperture: _____

Focal Length: _____

Eyepiece/Magnification:

_____

Filter(s): _____

Finder

Observer: _____

Date: _____

Time: _____

Location: _____

Seeing: _____

Transparency: _____

Lunar Phase: _____

Conditions: _____

_____

EP: _____   Mag: _____

Filter: _____   FOV: _____

EP: _____   Mag: _____

Filter: _____   FOV: _____

Notes

_____

_____

_____

_____

_____

_____

_____

_____

_____

_____

_____

_____

Observation Log

Object: _____

Constellation: _____

Instrument: _____

_____

Aperture: _____

Focal Length: _____

Eyepiece/Magnification:

_____

Filter(s): _____

Finder

Observer: _____

Date: _____

Time: _____

Location: _____

Seeing: _____

Transparency: _____

Lunar Phase: _____

Conditions: _____

_____

EP: _____ Mag: _____

Filter: _____ FOV: _____

EP: _____ Mag: _____

Filter: _____ FOV: _____

Notes

Observation Log

Object: _____

Constellation: _____

Instrument: _____

_____

Aperture: _____

Focal Length: _____

Eyepiece/Magnification: ____

_____

Filter(s): _____

Finder

Observer: _____

Date: _____

Time: _____

Location: _____

Seeing: _____

Transparency: _____

Lunar Phase: _____

Conditions: _____

_____

EP: _____ Mag: _____

Filter: _____ FOV: _____

EP: _____ Mag: _____

Filter: _____ FOV: _____

Notes

_____

_____

_____

_____

_____

_____

_____

_____

_____

_____

_____

_____

Observation Log

Object: _____

Constellation: _____

Instrument: _____

_____

Aperture: _____

Focal Length: _____

Eyepiece/Magnification:

_____

Filter(s): _____

Finder

Observer: _____

Date: _____

Time: _____

Location: _____

Seeing: _____

Transparency: _____

Lunar Phase: _____

Conditions: _____

_____

EP: _____   Mag: _____

Filter: _____   FOV: _____

EP: _____   Mag: _____

Filter: _____   FOV: _____

Notes

_____

_____

_____

_____

_____

_____

_____

_____

_____

_____

_____

_____

Observation Log

Object: _____

Constellation: _____

Instrument: _____

_____

Aperture: _____

Focal Length: _____

Eyepiece/Magnification:

_____

Filter(s): _____

Finder

Observer: _____

Date: _____

Time: _____

Location: _____

Seeing: _____

Transparency: _____

Lunar Phase: _____

Conditions: _____

_____

EP: _____ Mag: _____

Filter: _____ FOV: _____

EP: _____ Mag: _____

Filter: _____ FOV: _____

Notes

_____
_____
_____
_____
_____
_____
_____
_____
_____
_____
_____
_____
_____

Observation Log

Object: _____

Constellation: _____

Instrument: _____

_____

Aperture: _____

Focal Length: _____

Eyepiece/Magnification:

_____

Filter(s): _____

Finder

Observer: _____

Date: _____

Time: _____

Location: _____

Seeing: _____

Transparency: _____

Lunar Phase: _____

Conditions: _____

_____

EP: _____   Mag: _____

Filter: _____   FOV: _____

EP: _____   Mag: _____

Filter: _____   FOV: _____

Notes

Observation Log

Object: _____

Constellation: _____

Instrument: _____

_____

Aperture: _____

Focal Length: _____

Eyepiece/Magnification:

_____

Filter(s): _____

Finder

Observer: _____

Date: _____

Time: _____

Location: _____

Seeing: _____

Transparency: _____

Lunar Phase: _____

Conditions: _____

_____

EP: _____ Mag: _____

Filter: _____ FOV: _____

EP: _____ Mag: _____

Filter: _____ FOV: _____

Notes

_____
_____
_____
_____
_____
_____
_____
_____
_____
_____
_____
_____
_____

Observation Log

Object: _____

Constellation: _____

Instrument: _____

_____

Aperture: _____

Focal Length: _____

Eyepiece/Magnification:

_____

Filter(s): _____

Finder

Observer: _____

Date: _____

Time: _____

Location: _____

Seeing: _____

Transparency: _____

Lunar Phase: _____

Conditions: _____

_____

EP: _____ Mag: _____

Filter: _____ FOV: _____

EP: _____ Mag: _____

Filter: _____ FOV: _____

Notes

Observation Log

Object: _____

Constellation: _____

Instrument: _____

_____

Aperture: _____

Focal Length: _____

Eyepiece/Magnification:

_____

Filter(s): _____

Observer: _____

Date: _____

Time: _____

Location: _____

Seeing: _____

Transparency: _____

Lunar Phase: _____

Conditions: _____

_____

Finder

EP: _____   Mag: _____

Filter: _____   FOV: _____

EP: _____   Mag: _____

Filter: _____   FOV: _____

Notes

_____
_____
_____
_____
_____
_____
_____
_____
_____
_____
_____

Observation Log

Object: _____

Constellation: _____

Instrument: _____

_____

Aperture: _____

Focal Length: _____

Eyepiece/Magnification:

_____

Filter(s): _____

Finder

Observer: _____

Date: _____

Time: _____

Location: _____

Seeing: _____

Transparency: _____

Lunar Phase: _____

Conditions: _____

_____

EP: _____   Mag: _____

Filter: _____   FOV: _____

EP: _____   Mag: _____

Filter: _____   FOV: _____

Notes

_____

_____

_____

_____

_____

_____

_____

_____

_____

_____

_____

Observation Log

Object: _____

Constellation: _____

Instrument: _____

_____

Aperture: _____

Focal Length: _____

Eyepiece/Magnification:

_____

Filter(s): _____

Finder

Observer: _____

Date: _____

Time: _____

Location: _____

Seeing: _____

Transparency: _____

Lunar Phase: _____

Conditions: _____

_____

EP: _____ Mag: _____

Filter: _____ FOV: _____

EP: _____ Mag: _____

Filter: _____ FOV: _____

Notes

Observation Log

Object: _____

Constellation: _____

Instrument: _____

_____

Aperture: _____

Focal Length: _____

Eyepiece/Magnification:

_____

Filter(s): _____

Finder

Observer: _____

Date: _____

Time: _____

Location: _____

Seeing: _____

Transparency: _____

Lunar Phase: _____

Conditions: _____

_____

EP: _____   Mag: _____

Filter: _____   FOV: _____

EP: _____   Mag: _____

Filter: _____   FOV: _____

Notes

_____
_____
_____
_____
_____
_____
_____
_____
_____
_____
_____

Observation Log

Object: _____

Constellation: _____

Instrument: _____

_____

Aperture: _____

Focal Length: _____

Eyepiece/Magnification:

_____

Filter(s): _____

Finder

Observer: _____

Date: _____

Time: _____

Location: _____

Seeing: _____

Transparency: _____

Lunar Phase: _____

Conditions: _____

_____

EP: _____   Mag: _____

Filter: _____   FOV: _____

EP: _____   Mag: _____

Filter: _____   FOV: _____

Notes

_____

_____

_____

_____

_____

_____

_____

_____

_____

_____

_____

_____

_____

Observation Log

Object: _____

Constellation: _____

Instrument: _____

_____

Aperture: _____

Focal Length: _____

Eyepiece/Magnification:

_____

Filter(s): _____

Finder

Observer: _____

Date: _____

Time: _____

Location: _____

Seeing: _____

Transparency: _____

Lunar Phase: _____

Conditions: _____

_____

EP: _____   Mag: _____

Filter: _____   FOV: _____

EP: _____   Mag: _____

Filter: _____   FOV: _____

Notes

_____

_____

_____

_____

_____

_____

_____

_____

_____

_____

_____

_____

Observation Log

Object: _____

Constellation: _____

Instrument: _____

_____

Aperture: _____

Focal Length: _____

Eyepiece/Magnification: _____

_____

Filter(s): _____

Finder

Observer: _____

Date: _____

Time: _____

Location: _____

Seeing: _____

Transparency: _____

Lunar Phase: _____

Conditions: _____

_____

EP: _____   Mag: _____

Filter: _____   FOV: _____

EP: _____   Mag: _____

Filter: _____   FOV: _____

Notes

_____

_____

_____

_____

_____

_____

_____

_____

_____

_____

_____

_____

_____

Observation Log

Object: _____

Constellation: _____

Instrument: _____

_____

Aperture: _____

Focal Length: _____

Eyepiece/Magnification:

_____

Filter(s): _____

Observer: _____

Date: _____

Time: _____

Location: _____

Seeing: _____

Transparency: _____

Lunar Phase: _____

Conditions: _____

_____

Finder

EP: _____   Mag: _____

Filter: _____   FOV: _____

EP: _____   Mag: _____

Filter: _____   FOV: _____

Notes

_____
_____
_____
_____
_____
_____
_____
_____
_____
_____
_____
_____

★ Notes and Reflections ★

Made in the USA
Columbia, SC
29 May 2020